For all the "nuisance" geese and other
underloved birds of Brooklyn
—L.N.

For book pals Jen and Jen
—A.H.

Text copyright © 2025 by Lela Nargi
Jacket art and interior illustrations copyright © 2025 by Anne Hunter

All rights reserved. Published in the United States by Random House Studio, an imprint of
Random House Children's Books, a division of Penguin Random House LLC,
1745 Broadway, New York, NY 10019.

Random House Studio with colophon is a registered trademark of Penguin Random House LLC.

Visit us on the Web! rhcbooks.com

Educators and librarians, for a variety of teaching tools, visit us at RHTeachersLibrarians.com

Library of Congress Cataloging-in-Publication Data is available upon request.
ISBN 978-0-593-64755-4 (trade)
ISBN 978-0-593-64756-1 (lib. bdg.)
ISBN 978-0-593-64757-8 (ebook)

The artist used mixed media to create the illustrations for this book.
The text of this book is set in 17-point Garamond Premier Pro Medium.
Interior design by Paula Baver

MANUFACTURED IN CHINA
10 9 8 7 6 5 4 3 2 1
First Edition

the lonely
goose

WRITTEN BY

lela nargi

ILLUSTRATED BY

anne hunter

RANDOM HOUSE STUDIO ⌂ NEW YORK

*H*ere is how he won her over:

He was the best dancer with the most
powerful moves.
He hissed loudly and bravely.
He had a long, sleek neck and a fine bill.
When he called her, she swam nearer.

Out on the water they
bowed their heads together.

On land she stood beside him
as he preened his feathers.

She followed him as the flock
grazed in tender grasses.

At night they slept snuggled together.

He kept guard as she made her nest.

He honked and flapped if anyone came too near.

When their work was done, they nibbled seeds

and pond flowers. They swam and bathed.

The world was the two of them.

Day after day she laid an egg in her nest until there were six.

She settled her body on top. She tucked in
her head for the long wait.

She muttered gently to her growing babies.
They squeaked and muttered back.

The parents took their goslings to the pond.

They showed them where to forage for the
best tasty plants.

They honked to keep them out of danger.

They taught them to fly.

And when the time came, they flew south
all together with their flock.

Year after year the two returned
to the same nesting spot.
They raised more babies.
They swam and preened and
snuggled, just like always.

One spring he could not fly.

He could not preen.

He could not eat.

He closed his eyes on the bank of the pond.

For days she sat beside his body.

Her sisters came to sit with her.

She did not fly.

She did not preen.

She did not eat.

She stayed behind when the flock flew south.

She sat as the weather turned gray and the sky got cold.

The world was just her now.

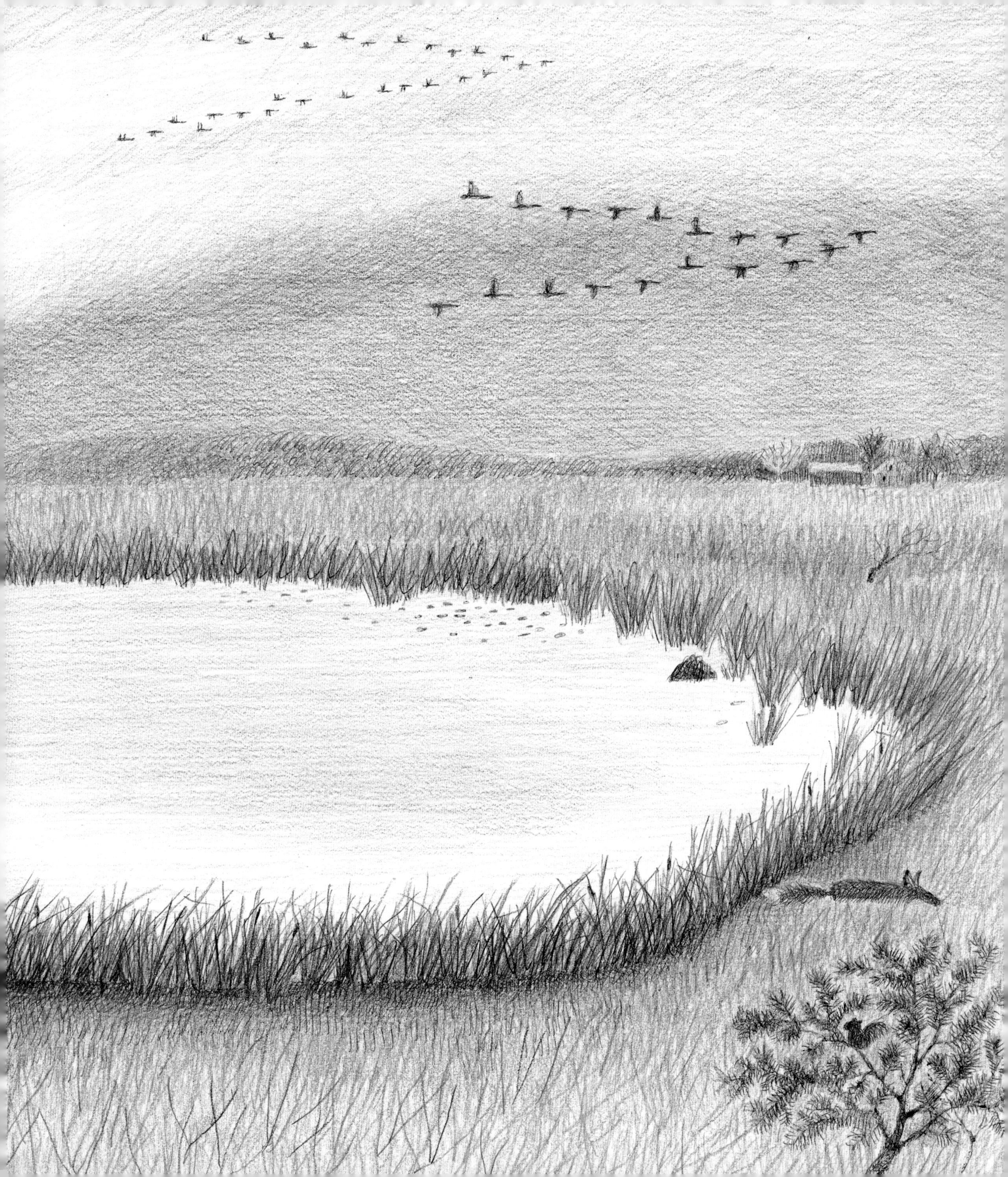

She sat as the snow melted and the ice
on the pond became water again.

She watched as the flock returned.

She saw her sisters.

She saw her daughters.

Little by little she felt hungry again.

She noticed who had come back this year.

And who had not.

And who for the first time
was all alone. Just like her.

He did not have to try to win her over.

He did not have to dance or hiss.

Little by little she came to stand
beside him as he preened his feathers.
She followed him as he grazed
in tender grasses.

In the middle of the pond they bowed their heads toward each other.

They slept snuggled together.

He kept guard as she made her nest and the days grew longer, brighter. He honked and flapped if anyone came too near.

When their work was done, they nibbled reeds and blueberries. They swam and bathed.

And now the world was the two of them.

More About Canada Geese

Canada geese can live for thirty years or more in the wild. They usually mate for life. They raise many broods of goslings together over the years.

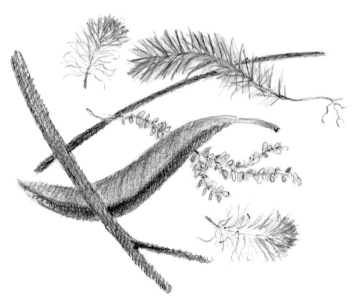

A female goose finds a quiet place to build her nest. She digs a groove in the dirt. She fills this with mosses, lichens, leaves, twigs, and her own soft down feathers. A gosling can hear its mother while it is growing in its egg and "talk" back to her.

The male takes his job as a protector very seriously. He will go to great lengths to chase predators away from his family.

The adults share parenting duties after their brood hatches. But an older sister might help her parents care for the new babies. The adult geese also might bring their babies into a larger group, called a crèche. There, several adults will look after many of the flock's goslings.

It's time to teach the goslings to fly when they are two or three months old. Parents show babies how to run down a hill and flap their wings. Off they go!

Mated geese will greet each
other loudly when they are
reunited. They communicate with
head, tail, and body movements
and with their voices. They have
thirteen different calls.

A goose will mourn for as long as
two years if its mate dies. Other geese
might keep that goose company for
a while. But then they will migrate,
while he or she will stay behind.
A goose without a mate will often
find a new partner—likely one who
also was recently widowed.

Lead in our environment is a huge threat to geese and all wild birds. Lead is a heavy metal. It gets into water, air, and soil when we burn oil and gas to drive our cars and use our machines. Geese might drink lead-poisoned water or eat plants that are contaminated with it. They are also losing habitat as humans replace wetlands with farms, roads, and buildings. That means there are fewer places for geese to raise their young.

Canada geese are not endangered like the California condor or the Hawaiian duck. There are lots of these geese around. But their lives are still affected by the challenges humans put in their path.

SELECTED SOURCES

"About Canada Geese," Citizens for the Preservation of Wildlife, preservewildlife.com/canada-geese.html

"All About Birds: Canada Goose," Cornell Lab of Ornithology, allaboutbirds.org/guide/Canada_Goose/overview

"Canada Geese Courtship," John I UK, 2014, youtube.com/watch?v=GMWUQidJvDk

"Guide to North American Birds: Canada Goose," Audubon Society, audubon.org/field-guide/bird/canada-goose

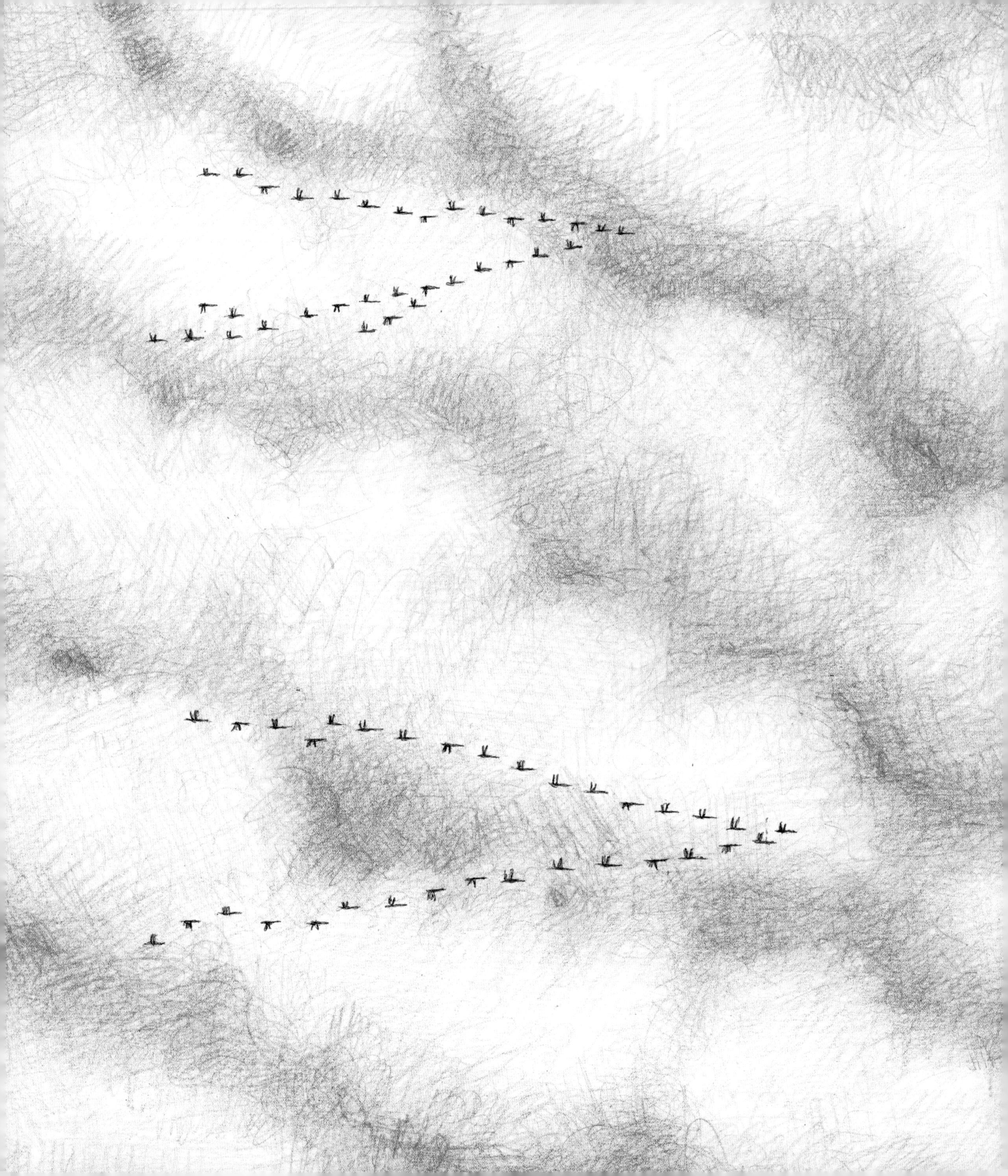